Benigno Nunez Novo
Maria Luiza Nunez Novo

Intergenerational dermatoglyphics in dizygotic twins and their parents

Benigno Nunez Novo
Maria Luiza Nunez Novo

Intergenerational dermatoglyphics in dizygotic twins and their parents

Dermatoglyphics and anthropometry in twins

ScienciaScripts

Imprint

Any brand names and product names mentioned in this book are subject to trademark, brand or patent protection and are trademarks or registered trademarks of their respective holders. The use of brand names, product names, common names, trade names, product descriptions etc. even without a particular marking in this work is in no way to be construed to mean that such names may be regarded as unrestricted in respect of trademark and brand protection legislation and could thus be used by anyone.

Cover image: www.ingimage.com

This book is a translation from the original published under ISBN 978-613-9-61533-9.

Publisher:
Sciencia Scripts
is a trademark of
Dodo Books Indian Ocean Ltd. and OmniScriptum S.R.L publishing group

120 High Road, East Finchley, London, N2 9ED, United Kingdom
Str. Armeneasca 28/1, office 1, Chisinau MD-2012, Republic of Moldova, Europe
Printed at: see last page
ISBN: 978-620-7-85850-7

Table of contents:

DERMATOGLYPHICS AND ANTHROPOMETRY
INTERGENERATIONAL IN TWINS
DIZYGOTIC AND THEIR PARENTS
MARIA LUIZA NÚÑEZ NOVO
BENIGNO NÚÑEZ NOVO

First of all, to God for his willingness to do this work for us.

Finally, we would like to pay a special tribute to our twins Ramón and Guilherne, to the people who encouraged us to do this work, to the friends we made while doing it and to everyone who helped make it happen.

INTRODUCTION

The personal characteristics of each individual are engraved on the human body in various ways, creating fertile ground for science to exploit these characteristics for the most diverse purposes, Multihull.

In studies with twins, the importance of the genotype in determining qualitative traits was evaluated, where the similarity of dizygotic pairs on average was presented by successively generated pairs, as they originated from different gametes produced by the genitors. Quantitative traits are determined by environmental factors, Beiguelman.

The genetic factor (qualitative or quantitative) manifests its totality when it finds favorable external conditions for its development, and when there is a lack of genetic information for a given characteristic, it does not develop, despite optimal external influences Filin.

Inheritance is a question that has intrigued the scientific community almost since the beginning of the 19th century, when the study of genes began with the classification and denomination of genetics. Batson's predecessor Mendel went so far as to identify recessive and dominant genes, and from these findings it was possible, by investigating the existence of semi-dominants, to discover that the issues surrounding genetics would require many more systematic and empirical studies to make satisfactory progress.

A person's development is the result of the interaction between heredity and the environment because the influence of hereditary factors and the environment on the development of certain traits is not equal. Some traits are subject to the action of the genotype; technically, heritability can only be interpreted as the process of transferring characteristics from one generation to the next, such as the color of the eyes of parents and children. The study of the degree of influence of heredity and the environment is the important objective of the genetic characteristic of a person.

The general laws of heredity, which are established between animals, plants and microorganisms, also extend to man. This once again testifies to man's biological unity with the entire organic world. DNA molecules, the nucleus of which is made up of concentrated chromosomes, serve as effective carriers of heredity at the different levels of organized life. The unit of heredity is the gene. In humans, 46 chromosomes (23 pairs) contain genes where all the information about heredity is concentrated. Thanks to the creation of a well-articulated and high-quality language, man differs from his ancestors, despite the community and union with nature. Man not only identifies himself by his biological particularities, but also constitutes a social being, as a member of a particular society. In order to better understand the correlation between the environment and heredity, it is very important to bear in mind that the progress of the human being, as a social creature, is not determined by heredity alone. Biological

heredity, related to the transmission of genetic structures, indicates that man has a

social continuity to be transmitted through education. Nowadays, taking into account

the biological and social particularities of a person, a series of methods for studying a

person's heredity have been developed and are widely used.

Part 1

DERMATOGLYPHICS, TWINS AND

ANTHROPOMETRY

CHAPTER I

DERMATOGLIPHY

Even before birth, human beings are susceptible to both genetic and sociocultural inheritance. In the case of genetic inheritance, there is little possibility of interference, unlike the field of sociocultural inheritance, i.e. the way in which the pregnancy is conducted, the tranquillity or disturbance of the pregnancy, nutrition and the emotional state of the pregnant woman, the formation and participation of the family nucleus, are factors that can be worked on and developed from the perspective of analyzing sociocultural influence, in which health and education professionals can also contribute and have been playing an important role in encouraging prenatal examination and monitoring pregnancy from various points of view.

The genetic characteristics determined by the hereditary factor determine to a considerable degree physical development, the formation of motor qualities, the body's aerobic and anaerobic performance and the level of increase in functional possibilities under the influence of sports training, in addition to the importance of the influence of the environment and human performance, contributing to whether or not genetic potential is achieved. It is important to emphasize that the use of prior knowledge of genetic abilities and tendencies, combined with the contribution of the environment, would help both in determining talent and in its development.

Heritability can be attributed to genetic causes, and there are two types of heritability: general heritability which represents the total contribution of genetic factors; and the restricted heritability resulting from additive components of genetic variance, i.e. average effects of individual genes on a character, they are not distinguishable from each other (SOBRAL, 1988).

Identifying the existence of sporting talent is a fundamental support for the detection (search) of sporting talent, which corresponds to all the ways used to find a sufficiently large number of children and adolescents to be included in a long-term training program (TLP). Talent selection is important if individuals with apparent potential are to be admitted to long-term programs, so reducing the risk of detection errors is becoming increasingly important.

Estimating the relative importance of heritability in determining the phenotype is important because it indicates the degree of influence that training can exert in relation to specific aspects, such as aptitudes and abilities.

By affirming the importance of heritability as a considerable variable in the dynamics of sports training in a broad sense, specific analyses can contribute directly to the planning stages and the establishment of programs in relation to physical activity practices.

Biologically, the study of the dynamics of this process has advanced sufficiently to produce information and present the theoretical construct that satisfies a large part of the understanding of this process.

The analysis of Dermatoglyphic and Anthropometric characteristics presents an advance in the use of techniques to identify genetic load, as well as serving as support for the identification of physical fitness and as a possible identifier of sporting talent, making it possible to anticipate and plan sports training, especially those aimed at high performance that need to be carried out over the long term.

Through the study of fingerprints (Dermatoglyphics) it is possible to identify the genetic predispositions of individuals in terms of races, genders and some of their hereditary characteristics, Borges, Matsudo.

CHAPTER II

TWINS

The use of twin pairs in studies to assess the relative value of the genotype in determining the phenotype (twin method), Beiguelman. They use variables that act on both the monozygotic and dizygotic twins found here.

Dermatoglyphics and anthropometry have been used to help with this identification. One of the fundamental elements to be considered in this context is the issue of genetic variability, which allows for the appearance of extremes in the population, which correspond to favorable phenotypic combinations and some of which seem to be highly dependent on heredity, such as: weight, height and muscular strength, Filin.

The relevance of this study lies in its empirical contribution, which reinforces the idea that the environment, i.e. qualitative variables, have a major influence on an individual's performance in sports. The study with twins reveals the equivalence of the characteristics of heritability, leaving the differentiation of factors due to environmental interference.

In twin studies, the importance of the genotype in determining qualitative traits was evaluated, where the similarity of dizygotic pairs on average was presented by successively generated pairs, as they originated from different gametes produced by their genitors. Quantitative traits are determined by environmental factors, Beiguelman.

Dermatoglyphics is the science that studies the relief of the skin and the design of the fingertips, palms and soles and the designs of the fingertips. Dermatoglyphic indices are formed in the human being in the intrauterine state from the blastogenic (embryonic) extract, ectoderm in the first three months of development and do not change throughout life, creating a permanent mark capable of revealing information about the individual's potential in relation to certain activities, such as sports, Beiguelman.

In this evolutionary process, environmental and social conditions and lifestyle represent the cultural inheritance transmitted in the family through education, modeling and socio-economic conditions, which have a direct or indirect effect on the phenotypic characteristics of the twins, while biological inheritance is related to the influence from one generation to the next, associated with a gene or group of genes encoded in DNA, responsible for carrying the information of biological heredity, Haywood.

It is thus clear that the association between genetically inherited aspects and environmental stimulation is fundamental to the development of human activities, which has an interesting effect on the performance of any physical activity that requires quick reactions, speed and agility, balance, coordination and strength Galahue, Fernandes Filho.

There are currently a growing number of authors and scientific papers evaluating the influence of twins, both quantitatively and qualitatively. The existence of genetic influence on levels of anthropometry and dermatoglyphics has been a basis for possible differences in twins. We must identify the influence of heritability, which makes a total contribution to the development of appropriate genetic factors. It is very important that Physical Education professionals use instruments and tests that allow them to achieve their goals with certainty and consistency through family studies, Beiguelman.

Heritability represents the total contribution of genetic factors. Dermatoglyphic characteristics can be obtained between individuals with a degree of kinship such as siblings, parents and children, Borges.
Fingerprints are characterized by almost parallel lines that form configurations. It includes the type of drawing, the number of lines on the fingers (the number of ridges within the drawing), the complexity of the drawings and the total number of lines, Fernandes Filho.

The shape of the drawings is a qualitative characteristic, while the number of skin ridges (lines) within the drawing is a quantitative characteristic. Most authors distinguish three groups of drawings

Arch (most commonly seen on the II and III toes); cleat (usually present on the III and V toes) and Verticulum (with a higher incidence on the I and IV toes), with the arch being found more rarely than the cleat and verticulum.

The verticillum is found a little more often and the cleat is the most common Fernandes Filho design. He points out that the fingerprint model makes it possible to choose the sports specialist more appropriately, with a view to optimizing individual talent. This is an excellent way for teams to specify the position of their players during the game, knowing their potential performance beforehand. It has been classified as a method of studying relief in the field of medical science. Fingerprints are hereditarily determined structures   of structural multiformity   ,differentiated phylogenetically and anthropogenetically for the execution of complicated mechanical and tactile functions, which are distinguished by their individual incomparability, that is, they are universal genetic marks, Fernandes Filho. Dermatoglyphic patterns are established around the fourth quarter of fetal life and remain stable with age, meaning that post-natal development plays no role in dermatoglyphic variability (except in some pathological conditions) and has an advantage over other physical or physiological measures in humans, Fernandes Filho.

Fingerprints reveal, in their characteristics, the processes of speed and growth Fernandes Filho.

They also allow us to form a scheme of principles for associating fingerprints with functional manifestations: endurance, speed, strength, coordination and physical activities.

The diagnosis of an individual's potential is a fundamental problem, both in theory and in practice, and Dermatoglyphics is a simple and effective method for determining the abilities and possibilities of each individual, Silva Dantas.

Genetic interference and the stability of training responses over time allow for early diagnosis and prognosis of sporting success.

This stability of fitness variables is the ability to predict future values based on previous observations Silva Dantas.

CHAPTER III

ANTHROPOMETRY

Anthropometry is a science that produces the application of scientific methods of physical measurements on human beings, seeking to determine the differences between individuals and social groups in order to obtain information. It is a very old science, and like all old sciences it has followed different paths, and the diversity of paths is both positive and negative.

Dermatoglyphics: a method of classification in typoscopy, a branch of medical science that studies papillary reliefs, Beiguelman.

Fingerprints (ID): the reproduction of the dermal ridges that make up the distal phalanx of the fingers. Fingerprints are formed in the third month of fetal life, together with the nervous system of the blastogenic stratum in the echoderm.

They don't change throughout life, Fernandes Filho.

Number of lines (QL): this is characterized by the number of lines of the skin ridges within the drawing; this number is counted according to the line connecting the delta to the center of the drawing, without taking into account the first and last lines of the Matsudo ridge.

Phenotype: understood as the sum of the properties of the individual in question, at a given stage of their development. These properties are widely variable: they include various structural features, biochemical and physiological characteristics, as well as the individual's psychological properties.

The formation of phenotypic properties is a complex process that occurs through interaction between the genotype and environmental factors, Multihull.

Genotype: is the hereditary basis of the organism, i.e. the set of all the genes that the organism receives from its parents.

Fingerprints (ID): This is the type of drawing, the number of lines on the fingers - the number of ridges within the drawing - in other words, the summary complexity of the drawings and the total number of lines. Fingerprints are formed in the third month of fetal life together with the nervous system in the blastogenic stratum in the ectoderm. They don't change throughout life, Matsudo.

Arc "A": The absence of tri-radii or deltas made up of ridges that cross the digital pad, Fernandes Filho.

Fang (L): is characterized by the presence of the delta and is characterized by a design in which the ridges of skin begin at one end of the finger, curve distally towards the other end, but do not approach the one where they begin. The clip represents an open design, Matsudo.

The "W" verticle is the figure formed by the presence of two deltas in the drawing. It is seen as a closed figure, in which the central lines are arranged concentrically around the core of the drawing, Fitness and Performance, Fernandes Filho.

Part 2

PROTOCOLS AND FEATURES

DERMATOGLYPHICS, COLLECTION OF

FINGERPRINTS AND

ANTHROPOMETRY PROTOCOLS

CHAPTER I

PROTOCOLS AND CHARACTERISTICS

DERMATOGLYPHICS

The Dermatoglyphics, Cumins and Midlo protocol was used to obtain the fingerprints. This procedure was carried out using the proper collection form, using A4 paper and a fingerprint collection pad.

After the anthropometric measurements have been taken with a tape measure, the correlation indices will be calculated, with the value of "r" (closer to zero) being the lower the hereditary burden.

The other statistical approaches will be used to characterize the sample universe investigated according to the parameters characterized as mean and standard deviation.

The closer the "r" value is to 1, the higher the hereditary load and the lower the "r" value, the lower the genetic load (Pearson's protocol).

The genetic methods applied were the most diverse, trying to reconcile qualitative and quantitative information. Also by observing and linking genetic information that considers dizygotic twins, such as the exhaustive inscription of methodologies. These informales express interesting phenomena with marked inter-individual differences.

This area interprets the extent to which genetic variation is attributable to genetic factors transmitted within families. This phenomenon of variation is an undeniable and well-reported fact of the differences between subjects. It can be said that genetic effects govern inter-individual differences and consolidate the importance of the family in the lifestyle of twins.

The data was collected in the cities of Bom Jesus and Currais - PI, on the scheduled days under the responsibility of the author, who was assisted by the body composition specialist, the dermatoglyphics specialist (papilloscopist) and also by the assistants, in addition to following all the protocols mentioned.

(Anthropometric assessment and Dermatoglyphics). The arm and forearm measurements of the twins and their parents and the dermatoglyphic impressions were taken according to the experimental procedure described above.

A double-blind process was used to identify the types of digital drawings.

Fingerprints must be taken using A4 paper and a special fingerprint pad.

Anthropometric measurements of the arms and forearms of the twins and their parents were taken using a measuring tape.

To be included in the study, the sample subjects had to be dizygotic twins of both sexes.

Being vetoed by those in charge.

They had any clinical symptoms that would hinder their participation in data collection during the period of the study.

Those who do not agree to the terms of engagement with the researcher.

Do not wish to participate, as a volunteer, without return or financial advantage.

Thus, the performance of this study was observed through relationships between groups of twins and their intentionally chosen parents.

CHAPTER II
FINGERPRINTING

When collecting fingerprints, we used the pre-inked roller system, which has a high-strength handle and approximate dimensions of 15.0 cm x 6.5 cm x 40 cm. See figure below. We also used sheets of A4 paper to record the participants and soap and water to wash the participants' hands after they had finished taking the fingerprints.

Figure 1- Print collector

Digital

To obtain the fingerprints, we proceeded as follows.

The collector is applied to the distal phalanges of each finger of both hands, the twins and their parents, taking care that the phalanges are covered with the ink from the side of the valar surface and from the sides up to the nails, so that the entire surface to be printed is covered with an even layer of ink, for the phalanx to be printed, it must be squeezed very carefully, without moving it, turning the finger from one corner of the nail to the other, towards the side of the index finger.

Figure 2 - Fingerprint collection procedure

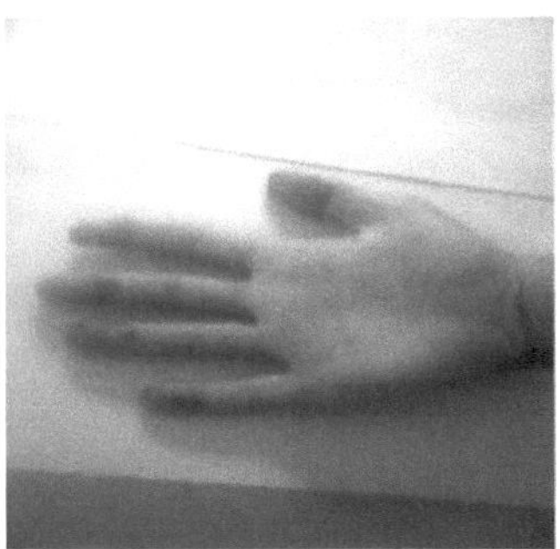

Source: authors

Once the fingerprints have been obtained, they are first read, using the following standard method. After the questionnaire has been administered, the fingerprints will be collected with the fingerprint roller on an A4 sheet of paper with their respective identifications.

Figure 3- Analysis and identification of dermatoglyphic patterns

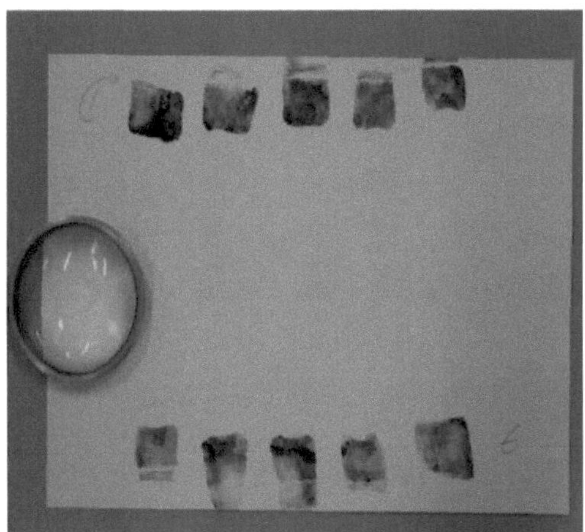

Source: Fernandes Filho. Training
Sport: Discovering Talent: CD ROM
2003

At the same time, the twins' arm and forearm anthropometric measurements were taken using a tape measure and placed on the back of the fingerprint collection sheet.

After collection, the fingerprints were divided into 3 types of drawings and classified by the number of deltas (Triradios) present in each fingerprint. Using a pre-inked roll and A4 paper, as well as the registration forms.

Figure 4- Bow Figure 5 - Clip Figure 6- Verticule

The basic standardized indices of fingerprints (ID) correspond to SQTL and D10:

The sum of the number of SQTL lines is the sum of the total number of lines on the 10 fingers.

In order to quantify the SQTL, it was necessary to count the lines, which is the process of determining the number between the nucleus and the deltas of the clips and verticils. The SQTL consists of counting all the papillary ridges reached by Galton's line, with the exception of the starting and finishing points. Galton's line is an imaginary line whose starting point is the element that classifies the delta, and its starting point is the element that classifies the delta.

The center of the nucleus, always located at the apex of a loop (Presilha) or a curved line, similar to the infection of a loop (Verticilo) is classified as follows:

Arch A: This type of design is distinguished by the absence of deltas in the fingerprint and its dermal ridges run transversely across the digital surface, see figure

Clip L - Characterized by the design of a delta, where its shape has a half-closed path, in which the dermal lines start at one end of the finger, bend distally towards the other, but do not approach the initial part of the lines, being considered an open design.

Verticillum W - Distinguished by the presence of two delta designs, where the shape has a closed characteristic, so that the central lines enter around the core of the design.

Figure 7- Count of lines between the core and delta of the fingerprints

Thread count - cleat - vertical thread count

Source: José Fernandes Filho (1997)

D10 corresponds to the sum of the drawings of the 10 fingers, where the minimum value is 0 and the maximum is 20. The value of 0 can be justified by the presence of the arc (A) which corresponds to the absence of deltas and the other values are represented by the numbers of the drawings which can be (L) = 1 delta and (W) =

2 deltas. The D10 values were found using the following formula: D10=ML+2(MW).

Source: Fernandes Filho (1997)

At the same time, the anthropometric measurements of the twins' arms and forearms were taken using a tape measure and placed on the back of the fingerprint collection sheet.

CHAPTER III

ANTHROPOMETRY PROTOCOL

The equipment used for anthropometric measurement was a sanny tape

measure.

And a mechanical scale with ruler model Mod. 110 Welmy[R]. As can be seen

in figure 8.

Equipment used in anthropometry.

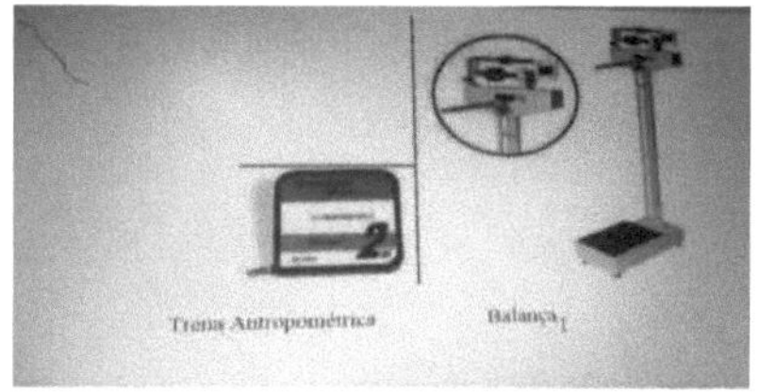

Source: Sanny and WelmyR

The following measurements were taken: body length - using a Sanny

measuring tape, accurate to 0.1 cm, acromial, radial-styloid. For this purpose, three

measurements were taken at each anatomical point by a single evaluator, in a rotational

manner. All the unilateral measurements were taken on the right hemibody, and the

median value was recorded, with the test-retest coefficient exceeding 0.95 for each of

the anatomical points, with a maximum measurement error of 5%. All subjects were

measured barefoot.

There are two methods of measuring body segments. One involves measuring

the vertical distance from the floor to a certain point on the body, with the person being

measured standing in the anatomical position (height) and the other involves measuring

the distance between a certain point on the body and another point on the body (length).

The height of the segments can also be determined by subtracting the body segments.

Part 3
EVALUATION, TREATMENT
STATISTICS AND RESULTS
CHAPTER I
EVALUATION

Arm length

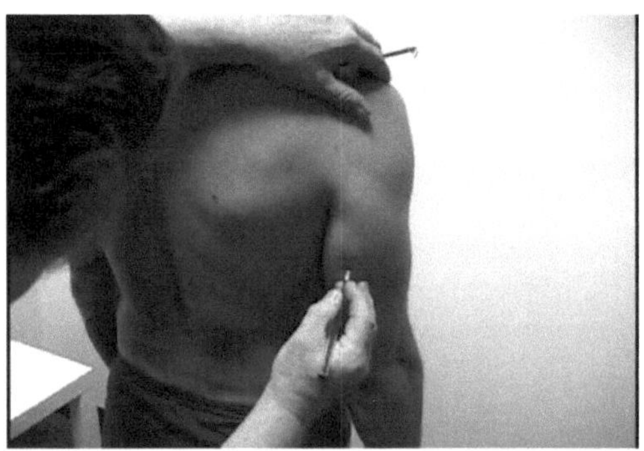

Definition

This is the distance between the acromial and radial points.

Protocol:

The subject should be standing in an orthostatic position with their palms facing their thighs. The assessor should place one of the caliper rods on the acromial point and the other rod on the radial point. It is not advisable to use a tape measure, as muscle volume, especially in the deltoid muscle, can interfere with the measurement

Forearm length

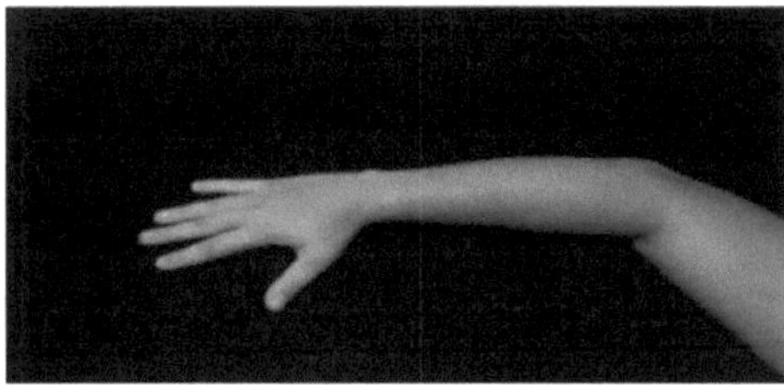

Source: author

Definition:

This is the distance between the radial point and the radial styloid.

Protocol:

The subject should be standing in an orthostatic position with their palms facing their thighs. The assessor should place one of the caliper rods on the radial point and the other rod on the radial styloid point.

This science produces the application of scientific methods of physical measurements on human beings, seeking to determine the differences between individuals and social groups in order to obtain information.

CHAPTER II
STATISTICAL TREATMENT

The statistical treatment used the discretionary values divided into mean, standard deviation and the variables applied. In order to guarantee their reliability, we applied Pearson's correlation test, which allowed for the non-parametric Mann-Whitney test and the Tukey test to check for significant differences between the groups ($p<0.01$).

It was applied to a small sample considering the variables under study, independent samples not presenting a normal or Gaugasian distribution. Descriptive statistics techniques were used to characterize the sample universe and to obtain discrete variables. Analysis of variance was applied to check whether or not there were differences between the means of the groups (father, mother, twins). Carried out at a significance level of ($p<0.05$).

CHAPTER III
RESULTS

This chapter presents the results obtained from the collection of data on

dermatoglyphics, anthropometric measurements and the questionnaire applied to the

twins and their parents, as well as their analysis using a statistical approach supported

by tables, graphs and histograms. For a better understanding, the following sequence

is followed: Dermatoglyphic variables of

Twins and parents; Anthropometric variables between groups.

Dermatoglyphic characteristics of twins and their parents

Graph 1 shows the descriptive values of the results for the number of twins

and fathers

And dermatoglyphic characteristics, where we can see that both parents and

children have a greater number of arches, with a tendency towards strength

activities, followed by the verticillum as motor coordination and finally the cleat

related to anaerobic activities. The trend of characteristics inherited from parents

prevails. The findings of this study show that heritability is linked to

Dermatoglyphic characteristics.

Dermatoglyphic characteristics of twins with their respective fathers - sum of

the total number of arches (A), cleats (L) and verticles (W).

Graph 1

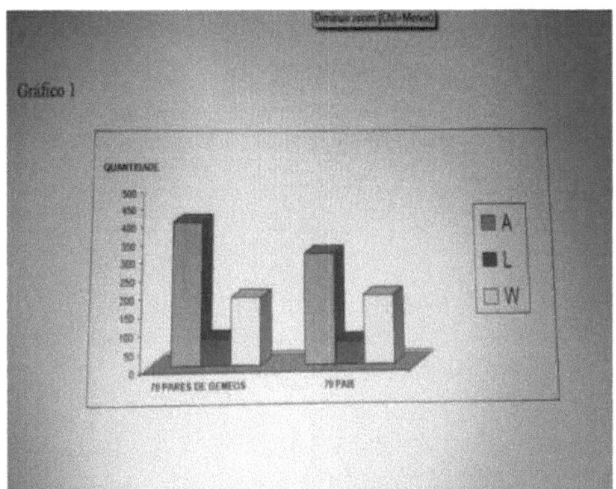

Table 1 shows the descriptive values of the dermatoglyphic characteristics Arch (A),

Loop (L) and Verticulum (W), in percentage of dizygotic twins and number of pairs

of dizygotic twins and number of pairs contacted in the study. Table 1

Dermatoglyphic characteristics of dizygotic twins in percentages of A, L and

W.

NAME	A	L	W
N	70	70	70
PERCENTAGE	9	61	30

Table 2 shows descriptive values
of the mean and standard deviation results in twins according to the dermatoglyphic
characteristics Arch (A), Loop (L) and Verticulum (W) in SQTL and D10 of
dizygotic twins.
Table 2

Mean and standard deviation of the Dermatoglyphic characteristics of Arch

(A), Loop (L) and Verticule (W) in SQTL and D10 of dizygotic twins.

NAME	SQTL	D10
N	70	70
AVERAGE	161,8	12,0
SD	43,61	4,17

Table 3 shows the descriptive values in percentages of the dermatoglyphic characteristics Bow, Cleat and Verticle and the number of parents taking part in this study.

Table 3 Dermatoglyphic characteristics of parents in percentages

NAME	A	L	W
N	70	70	70
PERCENTAGE	9	61	30

Table 4 shows the descriptive values of the mean, quantity and standard deviation results for the parents taking part in the research according to the dermatoglyphic characteristics Arch (A), Loop (L) and Verticulum (W) in SQTL and

D10.

Table 4

NAME	SQTL	D10
N	70	70
MEDIA	121,3	11,0
DP	11,3	0,12

Table 5 shows the descriptive data for the Arch (A), Loop (L) and Verticule

(W) of the male twins in percentages.

Table 5

Dermatoglyphic characteristics of male twins in Arch Press and Verticil

percentage.

NAME	A	L	W
N	11	11	11
PERCENTAGE	30	66	31

Table 6 shows the descriptive values of the mean and standard deviation results for dizygotic twins according to the dermatoglyphic characteristics Bow, Cleat and Verticle in SQTL and D10.

Table 6

Mean and standard deviation of the Dermatoglyphic Characteristics of D10 and SQTL male twins.

NAME	SQTL	D10
N	11	11
AVERAGE	134,9	12,7
DP	3,1	0,3

TABLE 7

Dermatoglyphic characteristics of female twins in percentage of Arc (A), Loop (L) and Verticulum (W).

NAME	A	L	W
N	14	14	14

PERCENTAGE	16	52	32

Table 8

Mean and Standard Deviation of Dermatoglyphic Characteristics of SQTL and D10 Female Twins.

NAME	SQTL	D10
N	14	14
PERCENTAGE	103,5	11,6
DP	2,3	1,2

TABLE 9

Dermatoglyphic characteristics of male and female twins in percentage of Arc (A), Loop (L) and Verticulum (W).

NAME	A	L	W
N	10	10	10

PERCENTAGE	8	66	26

Table 10

Mean and Standard Deviation of the Dermatoglyphic characteristics of male and female SQTL and D10 twins.

NAME	SQTL	D10
N	10	10
AVERAGE	129,5	12,1
DP	21,86	2,14

TABLE 11

Descriptive values of body segment lengths in cm of dizygotic twins and their parents.

CATEGORY	N	AVERAGE
FATHER	23	44.37

MÁE	34	41.77
GEMEO 1	34	37.75
GEMEO 2	34	35.37

Comparing measures Anthropometric measurements of the mother's arm with twins.
Graphs 1 and 2 show the descriptive values of the results relating to mothers

and twins.

Comparison of anthropometric arm measurements between mother and Twin

1 and 2.

Graph 2
Graph 3

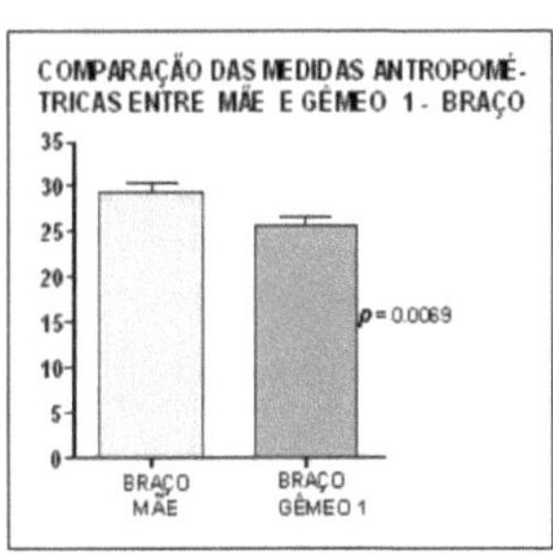

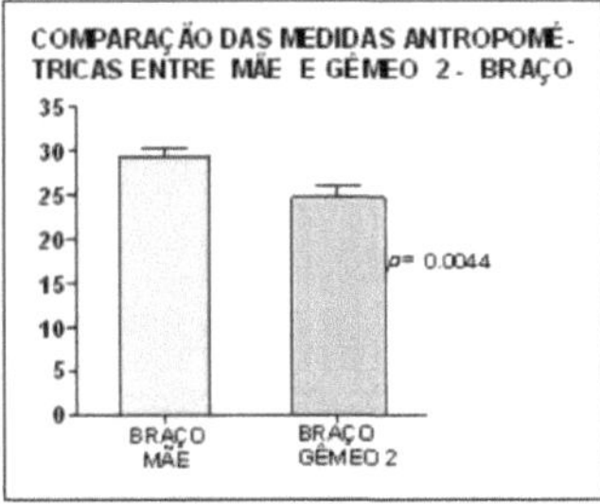

Table XII Correlation Measures
Anthropometric measurements (forearm) in twins and parents.

CATEGORY A	N	R	T	P - VALUE
	2	0.576	3.306	0.0032
PAI x G1	4	1	3	
	2	-	-	0.9488
PAI x G2	3	0.0142	0.065	
	3	-	-	0.9431
MÄE x G1	3	0.0129	0.072	
MAE x G2	33	-0.0945	-0.5287	0.6008
	3	0.754	5.754	P<0.000
G1 x G2	4	3	3	1

Table - XIII

Brago measurements of correlated parents Twins.

CATEGORY A	N	R	T	P - VALUE

PAI x G1	2 4	0.576 1	3.306 3	0.0032
PAI x G2	2 3	- 0.014 2	- 0.065	0.9488
MAE x G1	3	-	-	0.9431
	3	0.012	0.072	
		9		
MAE x G2	3	-	-	0.6008
	3	0.094	0.528	
		5	7	
G1 x G2		5.713	5.754	P<0.000
	3	1	3	1
	4			

Anthropometric data analysis
Arm and forearm of the twins and their parents.

The analysis of the anthropometric data (upper arm and forearm) of the twins and their parents revealed the prevalence of a hereditary burden in relation to the parents, the difference found being non-significant, as shown in the table correlating the three groups (father, mother and twins) with the upper arm and forearm measurements of both.

Correlated with the mother. The possibility of greater transfer of the father's characteristics to the twins means that more specific studies need to be carried out looking only at this part of the family (father versus twins).

The sample of twins is approximately 50% concentrated in the first grade age group. The genders are 44.8 for males and 55.7 for females.

This chapter presents the results found after analyzing the data collected, which were discussed in the light of the theoretical foundations that guided all the research work, demonstrating a direct link with the objectives of this thesis. Throughout the analysis of the dermatoglyphic indices found and their characteristics, Graph 1 shows that the predominant characteristic was a high level of Arches in both the twins and their parents, with the twins having a higher Arches index (A) than their parents. Corroborating this result are similar behaviors and findings in similar studies regarding the predominance of the Arc (A) design. These include the study carried out by Masjkey et al. in 2007, which found a higher incidence of Arches and lower rates of Verticillus compared to the control group, where they also identified a strong presence of ulnar clips (L) in interdigital regions II and III (66%) and a lower presence in region IV (18%). The appearance of genetic characteristics related to the mother (fingerprints) was also identified.

For the intergroup results, shown in tables 3,4,5,6,7,8,9,10, the characteristics of the group investigated are presented.

As for the Dermatoglyphic characteristics Loop (L), Arch (A) and Verticulum (W),

SQTL and D10, the behavior of the total number of lines on each finger of the parent

x twin group is similar to that found in the study carried out in Stockholm, Sweden,

by Nancy and L. Pedersen in 2005, showing that genetic influence is influenced by

the environment and that these can be conditioned with factors that influence the trait.

Influence on some cognitive abilities decreases with age. Human genetics has a bias

in its components. Dermatoglyphic studies reveal information that can make a

difference when it comes to choosing the type of physical exercise that will bring the

best results for you, because with Dermatoglyphics, it is possible to identify your

personal and non-transferable strengths and weaknesses and, from there, set up a

workout focused on your genetic potential where the results are faster and your

training is more efficient, which will motivate you to practice physical activity.

When looking at the results of the anthropometric variables, the study found

a significant difference (00.64) between the variables related to the mother and twins

in the male group. These   findings were similar to those of some studies by Silva

Dantas and Fernandes Filho, who also made comparisons between the genders, Filin,

Matsudo, Maia, and Fredericksen, Creistensen. The search for questions related to the

Dermatoglyphic and Anthropometric characteristics available in twins on genetics, purely and simply identifies an investigation to measure quantitative variables, which should indicate sufficient quantitative data to subsidize the research. These similarities can be justified by the fact that these individuals carry with them a series of phenotypic and genotypic manifestations of their own, which may be related to their state of health and consequently their physical development.

In relation to the questionnaire applied in the survey, the sample of twins shows a concentration of approximately 50% in the age range of 3 to 21 years. As far as gender is concerned, there is greater evidence of this in the 3 to 12 age group, corresponding to 24.28% of the sample for females and 17.4% for males, which is concentrated in the first grade age group. The gender represented in the twin sample was 44.28% for males and 55.71 for females.

Observing a questionnaire study and double-blind test on habitual physical activity in dizygotic (DZ) and monozygotic (MZ) twins from Stockholm, Sweden 2006. The sample consisted of 51 twin pairs of both sexes and different zygoties aged between 12 and 18 years. The mean values of the different indicators provided showed significant differences depending on physical activity and preferences during the week. It is therefore possible to highlight genetic effects in all the different expressions of this phenotype, showing that the genetic influence is influenced by the environment and these can be

conditioned with factors that influence swallowing. Influences on some cognitive

abilities decrease with age. Human genetics has a bias towards   dermatoglyphic

and anthropometric characteristics.

related, it can be concluded that the results of this study represent the construction of a

new era for modern and differentiated genetics through Dermatoglyphics and

Anthropometry within scientific standards that provide consistent practice, stratifying

and referring individuals for specific training in the area. Various studies are being

carried out which have concluded that Dermatoglyphics is an important factor in

determining the right profile for various sports in Brazil.

The statistical analysis used met the needs and provided tables and illustrative

graphs to better visualize the results, characterizing the profile of the dermatoglyphic

characteristics and the mean and standard deviation of the anthropometric

measurements of the twins and their parents. After analyzing the data and the results,

we will make them public.

The literature does not report any study of this nature, with a number of

dizygotic twins in Brazil in the proposed age range, so our work is extremely important

for the development of this study, where there is a gap in the national literature.

We would like to point out the immediate applicability of this study and its importance for the general population, by developing action strategies to prepare professional athletes through appropriate training and its characteristics. The cultural evolution of this study will serve as a benchmark for previously unknown professionals. This section is based on the conclusions drawn from the initial questions posed by the study, the general aim of which was to identify, compare, relate and evaluate dermatoglyphic and anthropometric characteristics in dizygotic twins aged between 2 and 55 and their parents.

It was concluded that twins show specific characteristics in anthropometric and edermatoglyphic variables   , with characteristics similar to studies carried out with twins in other studies.

This result represents the construction of a new era for modern and differentiated Physical Education, within scientific standards that provide, through knowledge, a practice consistent with its characteristics.

Finally, knowledge of these variables makes it possible to stratify and direct individuals towards specific training.

The results found in this study lead us to believe that further studies should be carried out in order to improve knowledge in this area.

Based on the observations made throughout this study, suggestions can be made for the continuation of this line of research in order to improve knowledge in this area even more: other studies should be carried out in other Brazilian states in order to increase the number of people evaluated.

Studies with the same boundaries and with monozygotic twins. To this end, it is recommended that the same study be carried out in groups other than the one in question, which would probably provide a broader view of the temporal scope of the phenomena, as well as investigating dizygotic twins and their parents in greater numbers.

FSC
www.fsc.org
MIX
Papier aus verantwortungsvollen Quellen
Paper from responsible sources
FSC® C105338